SUCCESS WITH KILLIFISH

EDWARD WARNER

 The Palmetto Publishing Company

Mini Pet Reference Series No. 1.

CONTENTS

Preface
1. Introduction — 1.
2. Accommodation — 3.
   - Aquariums — 3.
   - Water Condition — 4.
   - Lighting — 5.
   - Aeration & Filtration — 6.
3. Receiving Fish — 7.
4. Breeding — 8.
   - Mop Spawners — 8.
   - Peat Spawners — 9.
5. Raising the Fry — 13.
   - Food for Fry — 13.
   - Microworms — 16.
   - Algae Substitute — 16.
   - Freshly Hatched Brine Shrimps — 17.
   - Hatching out Shrimp Eggs — 17.
   - Live Adult Brine Shrimps — 19.
   - Grindal Worms — 20.
   - White Worms — 20.
   - Glass Worms — 22.
   - Mosquito larvae — 22.
   - Daphnia — 22.
   - Tubifex Worms — 23.
6. Accidental Hybridization — 25.
7. Questions & Answers — 26.
8. Killifish Species — 31.
   - Appendix A. Incubation Time — 36.
   - Appendix B. Incubation Solution — 39.
   - Appendix C. Spawning Mops — 40.
   - Color Plates — 41.

©1977 Edward Warner
Palmetto Publishing Co.
St. Petersburg, Florida.

All rights reserved. No part of this book may be reproduced in any manner without the prior written consent of the publisher.

Mini Pet Reference Series No. 1.

ISBN No. 0-915096-02-1

The main text of this book has been set in 10p Palatino by Photocomp, Inc.: using a computerized process, and printed by the Great Outdoors Press, St. Petersburg, Florida

Dedication

To my wife, Ruth, who has given up many things in life because of my fascination and devotion to tropical fish.

PREFACE

This book may well seem to be unusual in that it has no Acknowledgements or Bibliography. There is absolutely no material obtained from other sources, or gleaned from the writings of other authors. Any errors or unintentional false statements are my fault, and mine alone.

The background information that is the basis of this book, was obtained by personal involvement with tropical fish over a period of some 22 years, and specifically 12 years breeding Killifish, therefore I suggest that I am well qualified to write authoritatively on the subject.

My 'workshop' consists of over 200 tanks, all of which contain fish. Add this facility to the countless number of experiments that I have performed and distill it into the pages of this work and I am certain that the final result will prove of great help to other aquarists interested in pursuing this particular avenue of fishkeeping.

There are other books on the subject, but the majority of these are restricted to the old original information, with the same inaccuracies being carried on year after year. Also, when authors research information for books containing hundreds of species, these inaccuracies find their way into the new books. Anyway, that has been my experience.

Please remember that all my work was executed in my hatchery at Rockford, Illinois. It may well be that the methods that worked for me there will not always work for you in Great Britain, Florida or Oregon, or in some other part of the globe. Use this book as a guide. If you have been unsuccessful in breeding a certain fish, or raising the fry, start by using my methods, or incorporate part of mine with ideas of your own.

If this book has helped to further the cause of good fishkeeping then the purpose of this book will have been fulfilled and I will be content.

Edward Warner.
Jan. 1977.

1. INTRODUCTION

Most hobbyists who have become interested in Killifish are aquarists who have had some previous experience with other species of tropical fish, and who now wish to expand their knowledge by entering the fascinating realm of the 'Killies'. The purpose of this book is to guide the experienced aquarist and those aquarists who are venturing forth for the first time, along the path to successful Killifish husbandry and breeding.

When the author started keeping these attractive 'pond fish' there were about 180 known species, now there are over 750! It is obvious that a book of this nature could not possibly do justice to over 750 different fish, however, it is possible to cover those species usually available, and to proceed without too much difficulty. The Killifish field is growing so rapidly that any book published on the subject is more or less out-of-date before it reaches the aquarist.

A point that cannot be stressed too strongly is that the beginner should not be discouraged if he fails in his initial attempts to keep or breed these fascinating fish. Please do not start with the most difficult species, start with the easier species and then when you have had some success with them, graduate to the more difficult ones. Remember, a Boy Scout who has just completed a First Aid course is not ready for brain surgery.

This book has been designed to progress naturally through the various facets of keeping and breeding Killifish, it is a practical work, academic considerations have been more or less omitted intentionally since this information is available in other publications should it be needed. However, it is of interest to know something about the group of fish known as Killifish.

The name Killifish is derived from the Dutch word 'killi' meaning 'small pond'. Small ponds are the natural habitat of these fishes. Killifish have also been known in the past as Panchax.

The name is somewhat misleading, since it implies that these little fellows are fierce, and spend their time seeking and killing other members of the aquarium. This is, of course, just not true. Maybe some enterprising taxonomist will decide on a more appropriate name in the future.

Killifish belong to the family *Cyprinodontidae*, and they are often described as "egg laying tooth carps". They are closely related to the "live bearing tooth carps" which includes the *Goodeidae, Poeciliidae, Anablepidae*, etc. The main difference being the latter males have an external sex organ which is absent in the *Cyprinodontidae*.

The genera are:
- Aphaniops
- Aphanius
- Aphyosemion
- Aplocheilichthys
- Aplocheilus
- Chriopeops
- Cubanichthys
- Cynolebias
- Cyprinodon
- Epiplatys
- Fundulus
- Jordanella
- Leptolucania
- Nothobranchius
- Oryzias
- Pachypanchax
- Profundulus
- Pterolebias
- Rachovia
- Rivulus
- Valencia

A good representation from these are found under the heading 'Killifish Species'. Now some of the above are very rare and in some cases perhaps extinct. Others, like those of *Cyprinodon* genera, commonly called "Desert Pupfish" are on the Endangered Species List and collecting, harbouring and possession of these fishes is a violation of a Federal law with stiff penalties.

2. ACCOMMODATION

AQUARIUMS

Any of the standard sized aquariums are perfectly satisfactory for keeping Killifish for show, but for breeding purposes, small aquaria are advised. The reason for the small tank is to keep the fish in close proximity; every time the male fish turns, the female will be nearby, and the mop or peat waiting to receive the eggs. If a pair of Killifish are put into a 10 or 20 gallon tank, they will only have a passing relationship, and if the mop is not handy at the critical moment of spawning, then the number of eggs will be greatly reduced.

For the smaller Killifish, a 6 quart oblong bowl is sufficient. For the larger species, such as the Blue Gularis, use a 2½ gallon tank.

Freshly hatched fry can be placed in plastic dish pans for a week or two and then placed in 5 gallon or larger tanks to mature. When using plastic dish pans it is important to change the water daily, or at least, every other day, to prevent water pollution, which is the number one killer of fry.

When raising the fry to adults, use long tanks rather than deep ones. The fish seem to fare much better in this shape of tank, with the reduced depth of water and the consequent reduction in pressure. The air pumps are more efficient also since they do not have so much weight to lift.

The author does not supply any filtration or aeration to the small tanks containing the breeding pairs, but it must be remembered that he feeds only live foods. However, their water must be changed every week or two anyway, so in the author's opinion, filtration is unnecessary.

Killies are great jumpers, and they do not like strong light. For this reason hoods with lighting strips should not be used. The author uses screened tops made with aluminum screening stapled to a wood molding. (see sketch) This is an inexpensive device, easy to make, and has the virtue of being very effective.

Killifish will lose their color if they are put into aquaria that have light-colored gravel covering the base, therefore black or very dark gravel is recommended. Some individuals paint the outside surfaces of the back and sides of their tanks to give their Killies a greater sense of security. Killies will also lose their color if they are placed in bare tanks without any gravel. For this same reason, Killies housed in bowls for show purposes rarely do well.

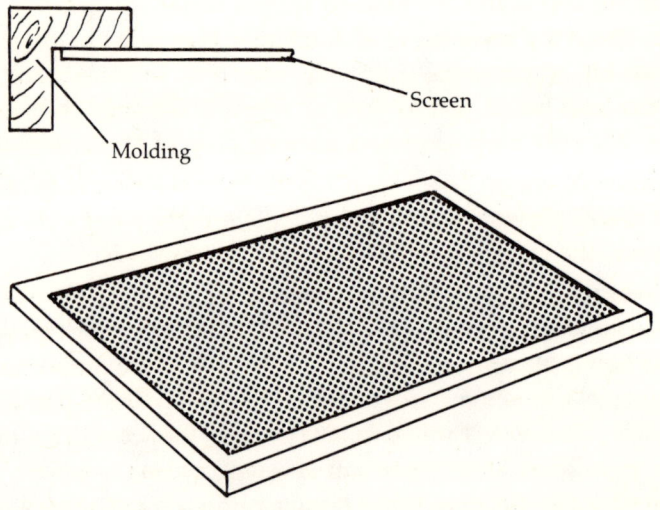

WATER CONDITION

Water conditions are really not as critical as many aquarists would have us believe. This is supported by the fact that the author has raised many Killies in hard alkaline water when books advise the use of soft acid water. But it is important to make all water changes gradually. Use about ¼ hard water and the rest soft water so that the resultant water is medium hard. This 'average' water allows the fish to be acclimated to any type of water easily.

Water drawn from a domestic faucet will undoubtedly contain traces of chlorine and copper ions. The chlorine is added to purify the water and the copper is absorbed from the piping. Chlorine can be removed by using 'Start Right'[1] or 'Clear Holdex'[1]. 'Blue Holdex' turns the water an ugly green when used with peat. Alternatively a few drops of pure sodium thiosulfate will do the same. Copper can be removed by filtering the water through activated carbon, but this should be done before any peat moss is added to the water, peat moss releases substances which are absorbed by the activated carbon and the result is that they effectively neutralize each other.

If the soft water is obtained from a water softener, do not worry about the presence of sodium ions, they will not affect the fish one bit. At times the author has used only soft water and the fish live and breed just as well as when it is mixed with hard water. With the mop spawners use one drop of Methylene Blue per gallon of water. The reason for the blue coloring is twofold: most Killies prefer the dim light, and since their eggs are light sensitive, the Methylene Blue serves a dual purpose.

Now this is very important. **ALWAYS** use a salt ratio of four (4) teaspoons of non-iodized salt per 10 gallons of water. This goes for the breeders, eggs and fry. Remember to add the proportional amount of salt when making water changes, either partial or complete. Persons who live in areas where they can draw their water from lakes should use salt at a ratio of one teaspoon per gallon of water because Velvet *(Oudinium)* is almost always present in this water and you will find it almost impossible to maintain the *Nothobranchius* species without the salt.

LIGHTING

It has already been stressed that Killifish prefer dim lighting conditions. It is an important factor, so no excuse is made for the repetition here. Hoods with strip lighting are not recommended, but if the aquarist does use such a method for lighting his tanks, the wattage must be of a very low order. It is preferable to use no special lighting, normal room illumination is sufficient.

1. When I mention a certain product name it is because I use it, although any SIMILAR product will do as well, of course. No commercialization is intended.
Ed Warner.

AERATION & FILTRATION

As stated previously, the author does not use any aeration or filtration in tanks containing breeding fish, the reason for this is that he only feeds live foods and changes the water frequently. The same applies to fry maintained in plastic dish pans. However, in tanks where one would normally house adults or fry, filtration must be suitable. Any filter that is strong enough to suck up the fry is obviously unsuitable, but a filter packed with charcoal and filter floss, even if it has a good rate of flow can be used successfully in tanks containing adults or half grown fish.

Undergravel filters can be used when there is no danger of losing fry. In fact, it would be safe to say that any method of filtration or aeration may be used with Killifish that would be used for other tropical fish. Remember, undergravel filters have a tendency to slowly acidify the water if water changes are not made periodically.

3. RECEIVING FISH

This procedure is sound for all tropical fish, including Killies. A great many fish are lost or disease organisms are brought into tanks when the proper method of releasing fish is not employed. By now, most hobbyists know they should not float their tightly sealed bags in their tanks.

When fish are obtained, either through the mail or direct from a dealer, use the following method. Open the top of the bag and float the bag in the tank where the fish are to be housed. Add about ¼ cup of tank water to the bag every hour or two. Continue to do this for 4 or 5 hours. It makes no difference what the pH or hardness of the water is in the bag, since you are slowly acclimating the fish to *your* water. Many hobbyists want to know the pH or hardness of your water when you ship them fish. What difference does it make? None if you follow this procedure. Now after the 4 or 5 hour period, dump the entire contents of the bag into a bowl or pail and immediately net the fish and put them into the tank from which the water came. In this way, the fish will not go into pH shock and will be ready for their new home. *UNDER NO CONDITION ADD THE DEALER'S WATER TO YOUR TANK*. If you do not already have disease, why contaminate your tank with some else's water? It is perhaps better not to feed the new fish for the first 24 hours; they need to become familiar with their surroundings and settle down a bit after being tossed around in a bag for some time. If you have an extra tank available for quarantine purposes, it would be advisable to use it. But in any event, use the method detailed above and your losses will be minimized.

4. BREEDING

While just keeping Killifish is a fascinating and very pleasurable pastime, breeding them is even more so. Obviously the better the stock, the better the offspring, so when you have a choice of breeding pairs, do select quality fish.

The male should have good body size, color, and fin structure. He should be robust and without defects.

The female should also have the same qualities as the male, with the exception of color, of course. Too often the author has seen pairs where the male is a beautiful specimen but the female is a runt. Such a combination cannot produce excellent offspring. Good quality fish cannot be guaranteed by 'group' or 'community' breeding where several pairs are kept together and bred in the same container. Therefore, it is advisable to keep breeding pairs separated.

In instances where the female is smaller than the male, try using breeding trios; that is two females to one male. This is a good approach anyway, because it will prevent the early death of one female particularly if the male is a hard driver. But do remember to use two **good** females.

The courtship of Killifish is quite similar to that of guppies which almost all hobbyists have seen at one time or another. The male will perform a 'dance' around the female with his fins extended and, at this time, his colors are most intense. With any cooperation at all from the female, spawning will soon follow. He will press the female into the spawning media until she expels an egg or two which are fertilized immediately. This activity may go on for an hour or two before the pair separates.

MOP SPAWNERS There are basically two types of mop spawners, the top mop and the bottom mop spawners. If the mop has a cork attached it will float, and it is known as a 'top mop'. If the cork is removed, the mop will sink and we have a bottom mop.

The preparation of the mops is described in Appendix C.

Nevertheless, other than the position of the mop, the breeders are handled in the same manner.

Prepare a small tank for breeding by adding one drop of Methylene Blue to each gallon of aquarium water, and put in the appropriate mop. The breeding trio can then be added. To avoid pollution of the water, be careful not to overfeed. This is important.

The mop spawners such as Roloffia and Aphyosemions produce adhesive eggs. Which is to say their eggs will stick to the mops and must be picked off by hand. Most mop spawners are bottom mop spawners and it is unlikely that any Killie will absolutely refuse to spawn in a bottom mop so if there is any doubt about how a fish should be spawned, use a bottom mop first and undoubtedly you will receive some eggs. Of course, their area should be free of excess traffic. There are some fish like *A. lineatus* and the *Bivittatum* complex which will give you a few eggs on a bottom mop but many more on a top mop.

Once the fish have finished spawning, remove a mop and gently squeeze the excess water from it, and look for the eggs. Do not be surprised at the great variation in the size of the eggs of different species.

The eggs are picked by hand from the mop and placed in small containers containing a hatching solution. Plastic Petri dishes are ideal for this purpose because the dishes can be stacked one on top of the other for space saving, and the eggs can be seen easily.

The hatching solution is described in Appendix B.

PEAT SPAWNERS

For Killifish that spawn in peat a different technique is required.

In the past many people have not enjoyed working with the peat spawners because of the trouble of preparing the peat. They had to cut, wash and boil the peat before it could be used. That is all in the past. Now there is a product called Jiffy 7 Peat Pellets. The Jiffy 7 are already ground to the right size and sterilized, they are autoclaved, which not only sterilizes them, but also compacts them so one pellet is actually a lot of peat. Also, one great advantage is that the pellets become water logged within five minutes, so your fish can start breeding in the peat almost immediately. Before the pellets, hobbyists often had to wait a day or two for the peat to sink to the bottom.

Now for the most important point of the Jiffy 7 Peat Pellets. There are four kinds of Jiffy 7 Peat Pellets. Numbers 700, 701, 702, and 703. Number 703 is *THE ONLY ONE TO USE*. Number 700 has the highest acidity and contains nutrients. We know nutrients are fertilizers. Number 701 has less acidity and less fertilizer and so on until we get to Number 703 which has a pH of 6.8 and no fertilizer. The author's water has a pH 7.2 so when the pellets are added to the water, the result is a pH of approximately 7.0, which is desirable. In this manner, it is possible to transfer the fish from a plain tank to the peat or vice versa without any pH shock. You see the greenhouses which use these pellets *WANT* fertilizer for use with their plants so Numbers 700 and 701 are the most commonly available. There are a thousand pellets in a box but you may go shares with someone to reduce the price and quantity. Jiffy 7 is a product of Grund, Norway.

(Remember, there is a lot of peat in one pellet.)

When using a 6 quart bowl for spawning, use one pellet but if the fish are what is termed 'Divers' like *C. whitei*, use two pellets. When using a 2½ gallon tank, use two pellets and again for the "Divers", add another pellet. Add one teaspoon non-iodized salt per 2½ gallons of water.

The object of the peat is to provide a spawning medium that will sink rapidly. In this respect the pellets are ideal because they quickly become water-logged and sink. The fish will not spawn in floating peat.

Standard garden peat can be used but it will be necessary to boil it first to help it sink and to sterilize it. The pellets are so much easier to use.

Having prepared the tank the breeding trios can now be added.

If possible, feed only live foods to reduce pollution. The

spawners should be kept in their tanks for a week or two, after which they are removed.

If the spawning has been successful, the peat laying on the bottom of the tank will contain fertilized eggs. To remove these use a nylon net to scoop up the peat and while it is still in the net, squeeze it to remove the excess water. It is possible to squeeze quite hard without doing any damage to the eggs. Now lay the damp peat out on newspapers. It is important that the newspaper selected should not have any colored advertisements or pictures that can come in contact with the peat, because the inks used are toxic to fish eggs.

The newspapers should be left on the floor containing the spread-out peat for 12-18 hours. By this time the outside of the peat should show signs of drying. Now place the peat and eggs into a plastic fish bag and add air, just the same as when transporting fish, and seal tightly. Use two rubber bands to seal the bags. Now label the bags with the name of the species you have in the bag. Place these bags in brown bags like those obtained from the grocery store. Label the brown bag with the date that its contents should be placed in water to hatch. Store these brown bags in a rather warm area.

In the Appendix, the incubation periods for quite a variety of different species will be found. Now when the day arrives for you to place the peat in water, use the following method. Take a container that has a large surface area, such as a plastic baby tub, fill with water to a depth of not more than two inches and add ½ teaspoon non-iodized salt. Remember to use salt with **ALL** Killies.

Do not add too much peat to a container. Use two or three containers if necessary. The reason for this is that the fry **MUST** come to the surface to get a gulp of air or they will become 'belly sliders'. If the surface is completely covered with peat it will not be possible for the fry to get this gulp of air. This is the most important point to remember when working with peat spawners. By the way, when the time has arrived to place the peat into water it need not be exact to the day. You may give or take a few days. This is one advantage of spawning fish in peat. They can be hatched at your convenience within a reasonable length of time, that is.

It is advisable to add a small quantity of Liquifry[2] to aid in the hatching of the eggs. A quantity of Microworms[3] may also be added to the container to induce hatching. This is called forced hatching but it can be used routinely because it also acts as a first food for the fry. After 24 hours, the fry should have emerged from their egg shells. Prepare a plastic dish pan with about one gallon of water and ½ teaspoon non-iodized salt. Place the container upon a shelf with good lighting. A student's desk lamp is useful here. Take a magnifying glass in order to see the small fry and a medicine dropper to scoop up the fry from the peat container and into the container of fresh water. This concludes the hatching process.

Now you must raise the fry successfully. Again, you must use some judgement. If breeding *N. guentheri*, you may have as many as 500 to 1,000 fry, in which case you must use more than one container to hold this amount of fry. First food requirements for many Killies are given along with the incubation time of the peat spawners in Appendix B.

---

2. When using Liquifry to force hatch eggs the process is accomplished because the Liquifry produces bacteria which erodes or attacks the cell wall allowing the fry to emerge.

3. When using Microworms to force hatch the process is accomplished by an increase in carbon dioxide given off by the Microworms. Of course, this is theory but it sure works.

5. RAISING THE FRY

Now that the fry are in their own containers without the peat, the job really starts. It will now show whether your efforts so far have been in vain or not. Some fish are easy to spawn, but raising the fry is quite another matter. Remember, the number one killer of fry is pollution with its resultant destructive bacteria. The number two killer is starvation.

Every day, or at least every other day, take another container, add the proportional amount of salt and water, and transfer the fry from the old water to this fresh water. Nothing stimulates growth of fish more than fresh water and a good diet. Most Killie fry can take freshly hatched shrimp immediately, but a few require infusoria at first. With a little experience you can look at the fry and tell if they are large enough to take the larger sized food particles. The shrimp which is not eaten will decay and cause pollution. The following section will explain about food selection for Killies. Included are some suggestions that will surprise even the more experienced hobbyist.

Do not be in a hurry to get the fry out of the dish pans and into regular aquariums. The daily or almost daily water changes are doing the fish a world of good. This they will not receive in their regular aquarium. Keep the fry in the dish pans for at least two to three weeks, maybe even longer; it is the best start they can receive. Also, when you do place your fry in aquariums, use the standard tanks and not the real deep tanks, as it appears the pressure in deeper tanks is not suitable.

FOOD FOR FRY

INFUSORIA

While it is true most Killie fry can take freshly hatched shrimp immediately, some will require infusoria for the first few days. The ease with which an infusoria culture can be maintained is reason enough for all of us to have one on hand should it be needed.

Infusoria are minute animals which live on decaying organic matter which is exposed to the air. They are usually microscopic in size and only a few species can be seen without magnification. What the aquarist desires in an infusoria culture is a great number of live animal bodies with a minimum of bacteria present. The fry of egglayers will eat the animal bodies but not the bacteria. It is bacterial action which causes the foul odor and scum accumulation which is to be found in most infusoria cultures. A good culture should be odor free. How then, do we prepare such an important food and yet eliminate these bad conditions? Simply go along with nature and provide the necessary conditions under which the infusoria will thrive.

First of all, we must have a glass container in which to start the culture. Do not use plastic. It is necessary for light to penetrate the container to produce some algae, and with plastic it is quite difficult to observe the culture. Beware of certain metals and wood. (danger of pH change and decomposition). A one gallon pickle or mustard jar will do fine. Next, we consider the water. One may use aged tap water, rain or creek water, but the best results are obtained by using old, but clean, aquarium water. Aquarium water already has some infusoria present. **DO NOT USE CHLORINATED WATER OR ANY SALT IN YOUR CULTURE.**

Into this water is placed a **SMALL** quantity of vegetation. Use a small amount of dried lettuce leaves, about 2 -3 inches in diameter. The infusoria culture should be placed where it can receive a good light to encourage the growth of algae, which helps feed the infusoria. This is all that is really necessary to start a culture, but for fast results the culture may be inoculated with a small quantity of infusoria water from a friend. Also, if certain organisms such as rotifers, paramecia, etc. are desired in your culture, it may be inoculated with cultures available from biological supply companies. Future cultures may be started by inoculations from your original culture, providing it has no odor. This culture should be allowed to stand at room temperature for at least a week before use. Once you have started an infusoria culture, you should be able to maintain it for years without much difficulty. Always leave your culture uncovered. To start a transfer culture, just take about ½ pint of your culture and add it to a clean jar with aquarium water and add a small piece of lettuce as before.

IMPORTANT — Do not put infusoria tablets directly into any fry tank as stated by some directions. There are three reasons for this:
1. We cannot regulate the number of infusoria.
2. This will bring about a lack of oxygen.
3. If something should go wrong, we will lose all of our fry.

Infusoria tablets are nothing more than dried vegetation. But if you desire to use them instead of dried lettuce leaves, add only ½ tablet to your infusoria culture. To purchase infusoria tablets is actually wasteful since the dried lettuce leaves do just as well.

The amount of infusoria that one feeds will depend upon the number of fry being fed and the size of the container that is housing the fry. With the plastic dish pans, about ¼ cup of culture once a day is sufficient. For the Killies that do need infusoria, it is only necessary for 2 to 3 days; by then they are large enough to take microworms or freshly hatched shrimp.

It is advisable to use liquid food in tubes specified as for egglayers in conjunction with the infusoria when feeding the fry for two reasons:
1. It will provide food for the fry.
2. It will provide food also for the infusoria.

These liquids contain milk and egg products and they are best refrigerated after opening the tubes, because they provide a good medium for the growth of bacteria also. As water is used from the infusoria culture, it should be replaced with aged water. If any odor develops in your infusoria culture, discard the complete culture and start another with everything new. Do not use any water from the initial odorous culture which obviously has a high bacterial content.

If for any reason it is impossible to feed infusoria when it is needed, frozen plankton may be substituted but only as an emergency measure. Plankton are minute organisms collected from the ocean and they are sold in many pet stores.

MICROWORMS

Microworms are discussed next because, believe it or not, they are smaller than freshly hatched brine shrimp. Microworms were discovered by Mrs. Grindal of Sweden. Grindal worms are named in her honor.

To culture Microworms, simply take a plastic container of about one pint capacity. Poke some holes in the lid to allow air to reach the contents. Add about ½ inch of any pre-cooked pablum: For example, Gerber's Mixed cereal. Corn meal cereals have not proved very satisfactory for rapid growth of the Microworms. Add water to the pablum until quite wet and then inoculate with some Microworms obtained from someone else who maintains a culture. Cover with the holed lid and keep at room temperature. Store the cultre away from bright light as there are no worms which appreciate light. After about 4 days, the worms should be climbing up the sides of the container. Use a knife to collect them and feed to your fry. Do not take any worms from the medium as you will be adding medium also to the water containing your fry and the danger of pollution will then be enhanced. Using only the worms from the sides of the container eliminates the need to wash the worms before use. Microworms may be fed in conjunction with freshly hatched shrimp or as a step between infusoria and freshly hatched shrimp.

Microworms are useful to force hatch fish eggs which require help. This was discussed earlier, but it is one of the main reasons for maintaining a Microworm culture, especially as they are easy to grow and maintain. They contain no harmful parasites and are an excellent food for fry.

They will provide something to fall back on if you have trouble hatching out shrimp; as a culture gets about one month old and looks bad, start a new one with everything the same as listed before. Just inoculate with the worms from the side again.

ALGAE SUBSTITUTE

Very often a hobbyist is in need of algae. It may be for fish that like vegetation in their diet, or perhaps for raising newly hatched shrimp or daphnia to maturity.

Simply blenderize fresh lettuce with a small quantity of water at high speed for at least four minutes. You may freeze a portion for use later. Remember, algae are *living* plants and are not prone to pollution; but the lettuce substance is not living, so do not overfeed or pollution will result.

FRESHLY HATCHED BRINE SHRIMP

This is by far the greatest food for fish fry and should be fed as soon as possible. The greatest contribution to the aquarium hobby has been the discovery of the brine shrimp egg and the method of hatching it. Before discussing how to hatch out the brine shrimp, something should be said about the brine shrimp egg itself and how to store it.

First of all, brine shrimp eggs must be at least one year old before they will hatch. If you could find shrimp eggs that have been vacuum packed and sealed for 5 years or more, you would get a hatch that would be almost 100%. If you purchase eggs which give you a poor hatch it is because they are less than a year old and should not have been put on the market.

It is very common to get a good hatch of shrimps when first opening a new pack of eggs, and for the hatch to get worse every day thereafter. The reason for this is that the main enemy of shrimp eggs is moisture and with the humidity of most fish rooms, it is not long before the eggs have absorbed the moisture and in consequence, will not hatch. If you should own a frost-free freezer, you are really in luck. Store your brine shrimp eggs in your frost-free freezer with the top of their container open. As you need your eggs, take out what is required and return the balance to the freezer. The unit that makes your freezer frost-free does so by removing moisture and that is what it is doing to the eggs — keeping them moisture free. If you do not own a frost-free freezer, the only thing you can do is store your eggs in a place where the humidity is the lowest and surely that is not your fish room. Keep the cover very tight to keep out moisture.

HATCHING OUT SHRIMP EGGS

All the brine shrimp containers have directions on how to hatch their eggs. The only disagreement to be found with these is the amount of salt they say to use. Sometimes one would think that they sell salt also because they recommend much more salt than is necessary for most hatchings. Having used shrimp eggs from various sources, and using a consistent quantity of salt, the author has obtained excellent hatches. Sometimes with less than half the amount of salt supposedly necessary, without any differences in the hatching percentage of the eggs.

Use a one gallon pickle jar or similar container. Do not use plastic because light doesn't penetrate very well. Add luke warm water up to the neck of the jar. To each gallon of water, add 3½ tablespoons of non-iodized salt. It is not necessary to use aquarium salt as it is too expensive. Purex salt can be bought in 50 pound bags, for about 2 to 3 cents a pound, far cheaper than any other salt you can buy, and it is free of impurities. Add one teaspoon of brine shrimp eggs to each gallon of water. If you have a multiple tank setup and need more shrimp, use a 2 to 5 gallon container to hatch out your shrimp, but add **ONLY** one teaspoon of eggs per gallon of water.

Now the jar must be very well aerated. Without proper aeration, the eggs will not hatch. The eggs should be kept swirling around in the water. The eggs will hatch out depending upon the temperature. At 75 to 80 degrees, the eggs will hatch in 18 to 24 hours. At a temperature of less than 70 degrees the hatching will take 48 hours. In this case, you should have two shrimp hatchers going on alternate days or use aquarium heaters in your jars for daily hatches of shrimp.

After your shrimp hatchers have stood the required length of time for the shrimps to hatch, place them on top of a counter. Then cover them with a brown bag, leaving a clear space of about 3 inches from the bottom. It may be necessary to fold the bag a little to get this space. Leave the bag on for at least 7 minutes, this is the length of time it takes for the shrimp to collect on the bottom. The shrimp are phototrophic, meaning they are attracted to light; that is the reason for leaving the blank space on the bottom of the jar. The empty egg shells should float on top.

After 7 minutes or longer, syphon the shrimp from the bottom into a clean jar or pail. Then using a cotton net, run the shrimp water through the net into the first jar so you can use the same salt water again. Any shrimp caught in the net is reversed in plain water. Using a baster you can then feed all your fry at one time. The shrimp water can be used about 4 to 6 days before it should be replaced with new salt and water. As a matter of fact, each day will probably give you a better hatch until the water sours. You have to use a cotton net because the shrimp would go through a nylon net. Cotton nets are getting more difficult to find these days so you may have to make your own. Just take the frame of an old nylon net and sew in a piece of an old handkerchief or other cotton material. See — wives are needed in this hobby, too.

LIVE ADULT BRINE SHRIMP

If a hobbyist purchases a large quantity of live adult shrimp he will want to keep them alive as long as possible. To do this use as large a container as you have available. Add some alkali to the water, such as Sodium Carbonate, because brine shrimp prefer alkaline conditions. Now for the important part; that is the amount of salt you use. Your hydrometer should read 1.050. The saline content of the water in which you receive the shrimp is probably in the region of 1.025 to 1.030. The shrimp will live much longer if kept at 1.050.

If you want to raise freshly hatched shrimp to maturity, the hydrometer reading should be 1.030 to 1.040. Feed the shrimp the aforementioned blenderized lettuce and crushed brewers yeast tablets. This is the same food for the adult brine shrimp. Also, if possible encourage the growth of algae, it helps their diet. A strange feature of brine shrimp is that if the water level is high they are livebearers and if shallow, they lay eggs.

Brine shrimp is the number one food in the hobby at the present time, but it may not be long before we find a substitute. The areas where the shrimp and their eggs are found are disappearing. Pollution like that of the Great Salt Lake is the main reason, although many areas are being taken over by private enterprises for boat launches, recreation and industry. Some experts say we must find a substitute by 1985 or we will have one large problem on our hands.

GRINDAL WORMS

Grindal worms are much larger than Microworms but smaller than White worms. They are an excellent food for half grown and adult fish. They are cultured somewhat differently than White worms.

Take a container similar to a plastic shoe box, mix equal parts of black dirt and peat moss, fill the box to a depth of about 3 inches. Inoculate with some Grindal worms. Feed *daily* with a small quantity of pablum on top and sprinkle with water to make the pablum moist. Cover the container with plastic material like that of a thick plastic bag. Lay the cover directly on top of the culture. The worms will adhere to the plastic; simply rinse the cover in clean water to remove the worms, letting them fall into a container. Then using forceps or a baster you can feed all your fish at one time.

Eventually the culture will become overrun with 'lice like' organisms and then you will have to start another culture. It has been said that one can prevent these 'lice like' organisms from growing by providing a moat around the culture. (Something like placing the culture in the middle of a large candy dish and surrounding the lower portion with water.) It certainly is worth the effort to see if this method works. It sounds like a good idea but it has not been tried by the author because at the present time he does not maintain a Grindal worm culture.

To start a new culture, take some Grindal worms only and place on top of a new setup of the black dirt and peat moss and proceed as before. Remember no worms like light so exclude as much light as possible from your culture; this is true for all worm cultures. Whereas White worms prefer cooler temperatures, Grindal worms may be cultured at room temperature without harm.

WHITE WORMS

For the successful cultivation of White worms, the container is very important. Do not use metal or plastic containers for two reasons. They do not "breathe" like wood; they absorb no moisture. A solid wooden box is used, which is also hard to find, so you may have to make your own. A box measuring 6 x 12 x 18 works fine.

Add equal parts of black dirt and peat moss to a depth of about 4 inches. The purpose of the peat moss is to prevent caking of the dirt. Add your White worms to the mixture. Now, this is the most important part of raising White worms; that is, feeding them. Feed them every other day. About twice a week sprinkle some Soya flour on top. Soya flour is made from Soya beans and may be obtained at health food stores. Alternate feedings with Gerber's mixed cereal on top. Pablums with the high protein contents do not seem to be favored by the worms. Now comes the most important part. Make a solution of 1 gallon water and a tablespoon of Rapid Gro. Any product may be used which is high in phosphoric acid, potash and nitrogen. These products are obtained in garden shops as plant food. Now, using a bottle sprinkler, sprinkle the food on top of the culture ultil the food is wet. Sprinkle lightly, but again, using a wooden box is advantageous in case you sprinkle too much.

Cover the top with a piece of glass to keep out insects, because the insects will lay their eggs in the medium if given a chance. White worms are like other worms, so keep them in a dark, cool place; an unheated basement is best. If the worm atmosphere becomes very warm, as it may in July and August, it is best to place a small culture in a refrigerator in case you "lose" the other culture; it can be started up again by using the aforementioned procedure.

When you want to feed White worms to your fish use the following method: on the day you want to feed worms to your fish, feed and sprinkle in the morning. By afternoon the worms will have collected on top and can be easily removed with tweezers to a small portion of water. It really is not necessary to wash the worms. Using the tweezers again, remove as many worms as you want to a cup of water and go right down the line feeding all your fish at one time.

The worms will get caught in the fibers of spawning mops and they are hard to remove. To remedy this situation, feed White worms to breeding fish right after their eggs are picked off and leave the mops on top of the containers until the fish have eaten all of the worms. Then put the mops back in again.

GLASS WORMS

Glass worms are larvae and cannot be cultured. You must purchase them from a dealer in season. The food value is often questioned and many fish will eat dry food before they will eat the Glass worms, so there is not much comment to be made on this point.

MOSQUITO LARVAE

This is one of the most nutritious of foods for fish. Again, using a large surface container like that for Daphnia, fill with water to near the top. This will allow for evaporation.

To attract female mosquitoes to lay their eggs in your container, take a nylon stocking and place some freshly cut grass in it. Float the nylon stocking in your container, put it outdoors for 24 hours and then remove it. It is the smell of the freshly cut grass which attracts the mosquitoes — not the nylon stocking. The nylon is for easy removal of the grass. This also encourages the growth of infusoria which the larvae will feed upon.

The mosquito eggs are first noticed as rafts of eggs. These can also be picked and let hatch in an aquarium. Depending upon the temperature, mosquito larvae should develop in about a week or so. Using a nylon fish net, moving it in a figure of eight pattern, the larvae can be removed. Remember, if you feed more larvae to your fish than they can eat you will have a room full of mosquitoes. You are doing your neighbors and yourself a favor by collecting mosquito larvae. The female mosquitoes will find a place to deposit their eggs under any conditions. But all those larvae you collected will not become mosquitoes to overrun the neighborhood. So if your neighbors complain to you explain that you really are helping them and you are not raising mosquitoes to plague them.

DAPHNIA

Daphnia may be cultured outside using a tub, refrigerator liner or child's wading pool as a container. Daphnia have a high tolerance for salt and seem to do better with some salt added to their water. So you should throw in a handful of non-iodized.

Use some gravel and dirt on the bottom of the containers to encourage the growth of infusoria which Daphnia also feed upon. They will consume any algae that will appear but their main diet should consist of ground up lettuce and brewers yeast tablets, which are crushed up to make the immediately available for food.

Starter cultures can be obtained from farmers who have a trough setting out. If you raise large cultures of Daphnia, they may be frozen and fed to your fish in the winter. There are both pros and cons about the food value of Daphnia but it does increase the size of young fish and brings about conditioning for breeders, so they are of some value. Many hobbyists swear by them and without doubt any live food is good for fish; Daphnia at the least provide a welcome variation in the fish's diet.

TUBIFEX WORMS

The word 'Tubifex' is the most mispronounced word in the aquarium hobby. Many people call the worms Tubiflex, please remember there is no "L" in tubifex. Hobbyists who have been active with fish for many years use this incorrect pronunciation.

It is understood that these worms are being successfully cultured in California and other places, although the author has not tried it, in a short period of time many hobbyists will be culturing their own tubifex worms.

These worms are sometimes called sewerage worms because that is where they are found along with garbage dumps. Sorry, sanitary landfill sites. Because of the habitat where these worms are found they must be cleaned before use. That means inside as well as outside. Here is a method for speeding up this process. Place these worms in a container like a small plastic dish pan with water slowly running over them in a sink. Place a small piece of sliced raw potato in with them. The worms will inject the starch water which helps to eliminate their intestinal contents. Using the continuous slowing running water and potato, the worms can be fed to your fish after 24 hours.

Many people fear feeding tubifex worms to their fish because of the many disease organisms they carry. This is a valid argument but tubifex is probably the best food for conditioning breeders. Germans are very fond of tubifex and feed it more often than Americans do.

There is also a Mexican variety of tubifex which is redder, thinner and longer than the American variety. These worms are handled in the same way as ours. Tubifex like cool temperatures and should be kept in a refrigerator if the hobbyist is going to keep them for some time. Do not worry, they will not crawl out of the container. It is very important that their water be changed *daily*, otherwise they will die of their own pollution. It may well be that we are learning to cultivate these worms because that will give us a cleaner worm and with the work of the Environmental Protection Agency, some day these worms might be hard to find.

Like White worms, these worms will get caught in your spawning mops, so feed them to your fish only when the mops are out of the tanks. Tubifex worms will bury themselves and live in the gravel but they are removed easily by washing the gravel or by allowing catfish to seek them out and eat them.

6. ACCIDENTAL HYBRIDIZATION

Since nearly all *Aphyosemion* females look alike and almost all *Nothobranchius* females look alike, there is a considerable risk of accidental hybridization. Here are the areas where the hobbyist must be alert:

1. Use the same mop used previously with the same species. All it would take would be a hidden egg or two to cause accidental hybridization. If this is not possible, then rinse the mop in very hot water between uses.
2. All pans of fry must be labeled. When transferring the fry to clean water make sure no fry adhere to the net. The safest method would be to dip the net into hot water between the transferring of different species.
3. When netting out peat and eggs to place upon newspapers, use a different net for each species. Of all the areas to be careful, here lies the greatest hazard of accidental hybridization.
4. When cleaning out aquariums which have housed adult fish make sure the gravel has been rinsed in very hot water. The adults will lay eggs in the gravel and these will hatch if not destroyed.

7. QUESTIONS AND ANSWERS

Listed here are some of the more commonly asked questions and answers about Killies.

QUESTION: Can Killies be kept together?
ANSWER: Yes and no. While it is true Killies of different species may be kept together, the females of the same genus but different species cannot be told apart. Therefore, if you ever intend to spawn these fish the answer is no. The males of different species may be kept together if they are of compatible size.

QUESTION: Are Killies community fish?
ANSWER: To the same degree that an angel fish is a community fish. By this analogy is meant they are safe with fish like Tetras and Barbs because they cannot catch them. They are safe with fish of comparable size but they are not safe with guppies. They love to nip the guppies tails as many fish do.

QUESTION: Do Killies really only live one year?
ANSWER: No. This idea has been brought about by the term annual fish. The annual Killies die because they are found in small bodies of water and of course when all the water evaporates, the fish must die. But this does not happen in an aquarium. Killies have about the same life span as livebearers. The fish should live from 1½ to 2 years, but in this length of time you should have many young fish to fall back on. It is not unusual for the Blue gularis to live 4-5 years but this is exceptional.

QUESTION: What happens when my eggs do not hatch when they are supposed to?
ANSWER: Force hatch them. There are several ways to do this. Whether you are speaking of water incubated eggs or eggs in peat, they can be treated in the same

way. First add a small quantity of liquid tube food for egglayers to the eggs. If after 24 hours this does not bring about results, add Microworms to the eggs that are in solution. If this does not work the eggs will never hatch. Either they are not good or the incubation period has not been long enough. See a later question about restive eggs.

QUESTION: What diseases are Killies subject to get?
ANSWER: Generally speaking, Velvet, particularly with the *Nothobranchius* genus. Killies can get Ich, but it is quite rare. Velvet can be controlled with the liberal use of non-iodized salt, about 1 teaspoon per gallon of water, or with chemicals readily available at aquarium shops.

QUESTION: Which Killies should a beginner start with?
ANSWER: The easiest to feed, raise and spawn. For mop spawners try the *gardneri - scheeli* group of *Aphyosemions*. And for peat spawners, *N. guentheri* or *C. whitei*. These fish have relatively short incubation periods and they are prolific, so some success can be expected.

QUESTION: How do you ship Killies?
ANSWER: Take a regular 6 x 12 plastic fish bag and twist the two bottom corners and hold them by using masking tape. This will make a round bag when you add water, fish and air. Add a small quantity of water and the fish to the bag. Inflate the bag with air from the room by closing the bag sharply by adding air from your compressor. Do not blow into the bag; this is adding carbon dioxide instead of air. Seal the bag with two rubber bands. Remember, air is more important than the amount of water because as the fish consume the oxygen from the water, it is replaced with the oxygen from the air in the bag. Now take this bag and place it upside down inside another plastic bag that does not have the corners tied and again seal with two rubber bands. This is called "double bagging". The packaging depends

upon the outside temperature and where the fish are going. If it is extremely hot or cold, place the fish bag or bags inside a styrofoam box and then inside a corrugated box. If the temperature is moderate you can place the fish bags inside a regular corrugated box and insulate with styrofoam chips, vermiculite, or any insulating material that is used to insulate homes. Seal the boxes well with either gummed brown paper tape or masking tape. Label clearly with "Live Fish" on the box. Ship Airmail - Special Delivery. With any cooperation at all from the postal service — the box of fish should be delivered within 24 hours safe and sound. This refers to the continental United States, of course.

QUESTION: How often do Killies spawn?
ANSWER: Unlike most egglayers that lay a large amount of eggs at one time and then wait about 2 or 3 weeks to repeat the process, Killies lay a few eggs almost every day. That is why we collect the peat and eggs every week or two and check our spawning mops for eggs every other day or so.

QUESTION: Should I use salt with my Killies?
ANSWER: Yes. Although Killies will live without salt, undoubtedly they do better with salt. Use a salt ratio of 4 teaspoons of non-iodized salt per 10 gallons of water. If you have trouble with Velvet, use a ratio of 1 teaspoon salt per gallon. Use salt with the hatching solution, rearing pans and with the adult fish.

QUESTION: What is the best way to handle fry?
ANSWER: Whether we use a medicine dropper to remove the fry from Petri dishes which were water incubated or from peat placed in water, the fry are handled the same way. If you use plastic dish pans, add a gallon of water and ½ teaspoon non-iodized salt to the pans. Then add your fry, using a medicine dropper. Feed the appropriate food whether it be infusoria, Microworms or nauphli. Change the water daily if possible, or at least every other day. This is to avoid

pollution which is the number one killer of fry. Nothing increases the growth of fry more than fresh water and plenty of groceries. After two weeks or so the fry should be quite large, they can then be placed in 5 or 10 gallon tanks, whichever you have available.

QUESTION: What is meant by restive eggs?

ANSWER: By now it will have become obvious that we are spawning the annual fish in peat moss instead of the mud of nature. We pack away the peat and eggs until the right time arrives to place them in water.

What happens in nature if, for example, a fish has a 4 month incubation period and a heavy rain occurs after 2 months? Nature provides a safety valve. While some fry will emerge and be under developed, easy prey for other animals or worse yet become a belly slider, this area will again dry up killing all fry. But there will still be some eggs in the mud that did not hatch because they were not developed enough to hatch. This is the safety valve. When the normal 4 month rain arrives, they will hatch. A few eggs will take 6 months to hatch as a second safety valve. We can do the same thing. If a fish has a 4 month incubation period, place it in water and collect the fry. Re-dry the peat on newspapers and bag up for another month or two. You will collect more fry, sometimes more than the first time. Repeat the process again for the third time and collect more fry if you wish. These later developed eggs are called restive eggs.

To determine the incubation period of new species find out where they came from — their locality. Using charts that will give the dry period in this locality will tell you how long the eggs should be incubated. Some charts will give the time that the monsoon season arrives, but it is still easy to calculate from this time. All remaining time would be the dry season. That is why the incubation period may vary with the annuals from 3 to 9 months, a few even longer.

8. KILLIFISH

A selection of species suitable for the beginner, with short descriptive notes.

Aphanius mento

Likes hard alkaline water with 1 teaspoon non-iodized salt per gallon of water. Sometimes called the poor man's *nigripinnis*. Very prolific and the eggs hatch in only 6 days in warm atmosphere. Eggs very adhesive. Top mop spawner.

Aphyosemion australe

The lyretails. Two species, the original chocolate and the mutant golden variety. Both handled in the same way. Bottom mop gives most eggs. Eggs are light sensitive. Contrary to what others say this fish appreciates water changes. Eggs hatch in 12-15 days. Does not require old acid water as some books advocate.

A. bivittatum

Several species in this group. All handled in same manner. Top mop spawner. Rather fast developer but sexing out rather slowly. Good fish for beginner.

A. celiae

Very prolific, but a slow grower. Bottom mop.

A. christyi

Very similar to lujae. Prolific with bottom mop but again a slow grower.

A. fallax

Spawn with bottom mop and lay eggs on top of peat for 5 weeks before placing in water.

A. filamentosum

A switch spawner. May be spawned with bottom mop or peat. Author prefers peat and incubates for 2 months. Good fish for beginner.

A. gardneri
> Many varieties now available. All bottom mop spawners and the first Killie for most hobbyists. Prolific and fairly rapid growth. Fry easy to care for.

A. melanopteron
> Top mop spawner. Very beautiful but not too hardy. Not a fish for a beginner.

A. puerzli
> Spawn with bottom mop. Lay eggs on peat for 5 weeks. A large Killie and good beginner's fish.

A. sjoestedti
> The Blue gularis. Largest of Killies. Spawn in bottom mop and lay eggs on top of peat for 6 weeks. Very prolific. Good beginner's fish.

A. sjoestedti
> Dwarf red gularis. Handled same way as Blue gularis but give eggs additional week on peat. This species has spots in middle of tail, red line through pectoral fins and is smaller than Blue gularis. Good beginner's fish and very colorful.

A. walkeri
> The orange variety is most colorful. Switch spawner author prefers peat and incubates eggs for 2 months. Easy fish to work with.

Aplocheilus lineatus
> Prolific, does not care where the mop is placed. Fairly rapid growth. Resembles a Pike. Large eggs. Good beginner's fish and a favorite fish of many old timers.

Austrofundulus transilus
> Bull dog looking. Spawns in peat with 4-5 month incubation. Males sometimes hard on femals.

Cynolebias antenori
> Easy fish to work with. Peat spawner. Sex ratio problem is common here. Shape similar to *C. whitei*.

C. bellottii
> The Argentine Pearl fish. Rather large fish. Favorite of

many. Peat spawner with 6 month's incubation. Fry not difficult to raise.

C. ladigesi

Not truly beautiful but easy to work with. Fry are quite small but they have a low mortality rate. Usually many more males than females.

C. nigripinnis

Very beautiful, always a good seller. Body has the appearance of jewels set against a black background. Peat spawner with 4-6 months incubation. This fish along with *N. rachovii* are the hobbyists' favorites.

C. whitei

This fish is large and most colorful. A "diver" in peat with 4-6 month incubation. Growth is extremely rapid with sexing possible at 2-3 months. An ideal fish for those wanting to try peat spawners for the first time. Fry are large, too.

Epiplatys dageti

The so-called fire mouth Killie. Really it is orange. Wild specimens can really be fire mouthed. Extremely prolific but the eggs are very small. The fry are small, too, but after a little infusoria the mortality rate is low. Many beginners have started with this fish as a mop spawner.

E. sheljuzhkoi

One of the most colorful of the *Epiplatys* family. Same as above and also a slow grower as is *dageti*.

Fundulus thierryi

One of the smallest Killies but nevertheless it is very pretty. Always a good seller. Fry are very small but growth is quite rapid with few fatalities.

F. grandis

A very large fish with full body. Green in color not very pretty, just large. The author had a pair that were 8 inches long.

Jordanella floridae

An American Killie. Found frequently in dealers'

tanks. Easily bred. So-called American Flag fish. Colorful.

Lucania goodei

Another American Killie. Quite colorful and easily bred. An overseas favorite and becoming more rare because they are losing their habitat. If you see it, buy it.

Nothobranchius guentheri

The first peat spawner for many. Success is almost 100% guaranteed. Two months incubation with peat. Fry are small but rapid growth. The author has developed a yellow strain that breeds true.

N. korthausae

Very beautiful. After incubation do not be in a hurry to discard the peat. Many fry take up to 3 days to hatch. Also, beware, most fry are found on the surface and easily missed if you are not careful.

N. palmquisti

Several varieties available. Good fish for beginner. Handle like *N. guentheri*. More red than *N. guentheri*. Colorful and easily raised.

N. rachovii

Everybody's favorite. On cover of many books. Easily spawned but raising the fry is difficult. Mortality rate is high, therefore this fish will never be available in great numbers.

Pachypanchax playfairi

Not too colorful but popular. Bottom mop, eggs are large, and so are the fry. Good fish for beginner. Easily spawned.

P. longipinnis

Very popular. Good color. Males' fins are split and some people think this is due to combat; but they are just natural. Easily bred, with fry no problem to raise.

Pterolebias peruensis

Long-time favorite. Six months incubation required. Large and colorful makes it worth the waiting. Fry easily raised with rapid growth.

Rachovia splendens
> Extremely prolific. Hatches of 1,000 or more are common. Not too colorful, but fun to have nevertheless. 4-5 month incubation. Mortality rate is low. Fry easily raised.

Rivulus cylindraceus
> One of the more colorful of the *Rivilus* group. Eggs are large and so are fry. Bottom mop. Good beginner's fish and usually available in the hobby.

R. harti
> One of the largest *Rivilus*. Same as above but not as colorful, orange ring on tail about all for color.

Roloffia species
> It is not for sure if this is a valid species now. Not described much in this book because they are all extremely slow growers, but also very beautiful. Bottom mop. Fry not difficult. Usually finicky eaters.

APPENDIX A

INCUBATION TIME

The following pages contain incubation times for some of the more common Killies. Since there are now more than 750 Killies and the author could not have worked with all of them these charts are representative. For instance, the information on Rivulus hartii and Rivulus cylindraceus could be used for most of the Rivulus species. Temperature also plays an important role on incubation. For the water incubated eggs, if they are kept cooler than 72 degrees F. one should add a couple of days; and in the case of peat stored eggs, one should add a couple of weeks if it is winter time or the eggs are stored in a cool room. Also, one should use these charts only as a guide, because what works for the author in Rockford, Ill. may not work for someone in Orlando, Florida. But it does serve as a place to start. One should experiment by separating the bag of peat and eggs into three parts. Try one part at say 3 months, another part at 4 months, and the last part at 5 months. If nothing hatches for you at 3 months and a few hatch out at 4 months, but at 5 months you get a good hatch, then 5 months is the time you will want to work with in the future. Incubation time will vary with individual fish within the same species, so find out which gives you the best results. Some species of Killies are what we call switch spawners, which means their eggs may be stored in peat or water. These particular fish are recorded in the chart as the way the author handles them, your method may vary. Do not argue with success. If you are happy with another method which produces good results, by all means continue it; but if you are not having success, try the methods in this book or a combination of both yours and the author's.

KEY:
PEAT = Whether the fish are spawned in peat or not.
FIRST FOOD = I = Infusoria; N = Nauphli, which is freshly hatched shrimp.
INCUBATION TIME = In number of months, M; or days, D.
LOTP = Lay on top of peat.
RATE OF GROWTH = S = Slow; M = Medium and F = Fast.

Species	Peat	First Food	Incubation Time	LOTP	Rate of Growth
Aphanius mento	No	N	12-20-D	No	M
Aphyosemion australe	No	N	12-20-D	No	M
A. batesi	No	N	12-20-D	No	M
A. bertholdi	No	N	12-20-D	No	M
A. bivittatum. (All species)	No	N	12-20-D	No	M
A. bualanum	No	N	12-20-D	No	M
A. celiae	No	N	12-20-D	No	S
A. christyi	No	N	12-20-D	No	S
A. cinnamomeum	No	N	12-20-D	No	S
A. cognatum	No	N	12-20-D	No	M
A. fallax	No	N	4 M	Yes	F
A. filamentosum	Yes	I & N	3 M	No	M
A. gardneri (All species)	No	N	12-20-D	No	M
A. gulare	No	N	2 M	Yes	M
A. lujae	No	N	12-20-D	No	S
A. puerzli	No	N	1½ M	Yes	M
A. santaisabellae	No	N	12-20-D	No	M
A. scheeli	No	N	12-20-D	No	M
A. sjoestedti (Blue gularis)	No	N	1½ M	Yes	F
A. striatum	No	N	12-20-D	No	S
A. walkeri	Yes	N	2 M	No	M
Aplocheilus lineatus	No	N	12-20-D	No	M
A. panchax	No	N	12-20-D	No	M
Austrofundulus dolichopterus	Yes	N	8 M	No	M

Species	Peat	First Food	Incubation Time	LOTP	Rate of Growth
A. transilus	Yes	N	6 M	No	M
Cynolebias adloffi	Yes	N	3 M	No	M
C. alexandri	Yes	N	4-5 M	No	M
C. antenori	Yes	N	6 M	No	F
C. bellottii	Yes	N	5-6 M	No	F
C. ladigesi	Yes	I	2-4 M	No	M
C. nigripinnis	Yes	I & N	5-6 M	No	F
C. whitei	Yes	N	5 M	No	F
Cyprinodon atrorus	No	N	12-20-D	No	M
Epiplatys dageti	No	I	12-20-D	No	S
E. sheljuzhkoi	No	I	12-20-D	No	S
Fundulus thierryi	Yes	I	3 M	No	M
F. grandis	No	N	12-20-D	No	M
Jordanella floridae	Yes	I	½ M	No	M
Nothobranchius furzeri	Yes	I	6-9 M	No	F
N. guentheri	Yes	I	1½ M	No	F
N. kirki	Yes	I	3 M	No	F
N. korthausae	Yes	I	3 M	No	M
N. melanospilus	Yes	I	3 M	No	F
N. neumanni	Yes	I	3 M	No	M
N. palmquisti (All)	Yes	I	3½ M	No	M
N. rachovii	Yes	I	6 M	No	F
Pachypanchax playfairi	No	N	12-20-D	No	M
Pseudepiplatys annulatus	No	I	12-20-D	No	S
Pterolebias longipinnis	Yes	N	6 M	No	M
P. maculipinnis	Yes	N	6 M	No	M
P. peruensis	Yes	N	9 M	No	M
P. zonatus	Yes	N	8-10 M	No	M
Rachovia brevis	Yes	I & N	6 M	No	M
R. hummelincki	Yes	I & N	6 M	No	M
Rivulus cylindraceus	No	N	12-20-D	No	M
R. harti	No	N	12-20-D	No	M
R. Riv. milesi	No	N	12-20-D	No	M
Roloffia species	No	I & N	12-20-D	No	S

APPENDIX B

INCUBATION SOLUTION FOR HATCHING EGGS

1 gallon water
½ teaspoon non-iodized salt
1 drop Aquari-Sol
5 drops methylene blue

The purpose of the methylene blue is twofold. First, it reduces the light which is important, because most Killie eggs are light sensitive. Second, if the eggs absorb the methylene blue and turn blue, discard them. Use a medicine dropper to remove them. They are either infertile or are fungused. Do not use more than 5 drops of the solution or all the eggs will turn blue. Methylene blue is not stable, so make up a fresh solution every 3 to 4 days. Methylene blue is not an anti-fungus agent. The Aquari-Sol is used as a fungicidal agent.

Other dyes like Acriflavine may be used instead of methylene blue, but in most cases they are too powerful for the beginner to use. Beginners have a habit of using much too strong a dye solution, so the methylene blue leaves them a little leeway.

The author recommends plastic Petri dishes to hold eggs and hatching solution. The dishes may be stacked atop each other cutting down on space. The dishes should be kept out of light as much as possible and checked daily. Any eggs which have absorbed the methylene blue should be discarded immediately using a medicine dropper.

APPENDIX C

SPAWNING MOPS

It is most important to select the proper material in which to make spawning mops. The ideal fabric is 100% nylon which is difficult to find today. Do not use Orlon or wool fibers. Nylon and Dacron are fine. Wool will rot in water and Dacron appears to be somewhat toxic toward fish.

Take a piece of cardboard, book or something similar about 10 inches in length and wrap the fabric around it lengthwise so you will have between 100 to 125 strands of mop after you cut the strands in two at the bottom. Make a small ball of fabric at the top and allow 2 strands to be free. This will enable you to place a cork on these 2 strands if you desire for the few top spawning Killies like Lineatus and Bivittatum species. Also, it makes it easy to hang up the mop when not in use.

If the fabric is darkly colored, leave it. If dyeing is necessary use a strong solution of methylene blue with some salt added to the water. The salt helps the dyeing process. Leave the mop in this solution overnight to dye the mop completely. Methylene blue is not stable so additional dyeing will be necessary after 4 to 6 months. The reason we dye the mops are twofold. First, the fish seem to prefer a darkened mop and secondly, it makes it much easier for us to find the fish eggs. There are some commercial mops available on the market which are quite coarse, but Killies do not care to use them for spawning eggs.

Do not use plastic fishing corks because they contain copper wire. Use plain corks which have a hole through them with either wood or plastic stops for attachment to the mops. By the way, those nickel corks now sell for 30c each.

Chocolate Lyretail. *(Aphyosemion australe)*.

Gold Lyretail. *(Aphyosemion australe)*.

Multicolor species. *(Aphyosemion bivittatum)*.

Aphyosemion bivittatum bitaeniatum.

Blue Gardneri species. *(Aphyosemion gardneri).*

Yellow Gardneri species. *(Aphyosemion gardneri).*

Misaje strain. *(Aphyosemion gardneri)*.

Aphyosemion mirabile traudeae.

Aphyosemion puerzli.

Blue Gularis. *(Aphyosemion sjoestedti).*

Dwarf Red Gularis. *(Aphyosemion sjoestedti)*.

Aphanius mento.

Cynolebias whitei.

Nothobranchius guentheri.

Nothobranchius korthausae.

Pachypanchax playfairi